Theory of Global Warming & Climate Change

Micro Model,

$$W = C\,(x)^{-1}$$

$$\theta = w\,(90\,/100)$$

$$Gw = \theta\,/\,66.02328T^{-1}\ {}^{\circ}C$$

Macro model,

$$GW = f(P)$$

$$= f\,(\,CO_2,\ O_2,\ Re,\ GHGo,\ AntiGHGo, TEc\,)$$

$$\Delta GW\,/\,\Delta P = h = \tan\theta$$

Mohammad Abu Shahadat

Theory of Global Warming

& Climate Change

by

Mohammad Abu Shahadat

Contents

About the Author

Works in 2019:

Theory of Global Warming , The Central Point of the Universe

Mohammad Abu Shahadat

Founder & Headteacher, Notre Damian School

Student, Welfare Economic System since 2013

Practitioner, Ecophysics

BSS (Hon's), MSS in (Economics), Jagannath University, Dhaka

Studied from 2002 to 2004 at Notre Dame College, Dhaka.

Studied from 1997 to 2002 at Delpara High School

Studied from 1992 to 1996 at MDC Model Institute

Born on 14 January 1987,Dhaka, Bangladesh

Father's name: Mohammad Badsha Mia

Mother's name: Anwara Begum

Spouse: Tanha Jerin Tammi

Son: Mohammad Shahriyar Shayan

Contact: www.facebook.com/MAbuShahadat

Email:mabu.shahadat@gmail.com

Who Will Benefit to Read This Book

Global Warming theory designed by Mohammad Abu Shahadat from Bangladesh is used to measure the global warming and how much CO_2 , CH_4, N_2O and other greenhouse gases are responsible for 1° Celsius global warming and how to minimize global warming by applying this model and how to reach the healthy dreamy world.

This book is helpful the students of PhD, Masters and Honours of Economics, Development Studies, Natural Disasters Management, Botany, Zoology, Geography, Physics, Life Science and Climate Change related subjects. This theory will be used by the researchers and scientists of various research center for climate issue. It will be also used by The World Bank, The United Nations, NASA etc for measuring Average Global Temperature.

Other students who want to gather knowledge about global warming and climate change.

The Author's Speech for The Readers

I, Mohammad Abu Shahadat, The Founder and Headteacher of Notre Damian School, declare that I always try to do something different and the work which nobody thinks or does before me I would like to do . I founded Welfare Economic System and Development in 2013. Since then to develop the welfare economic system I individually have been working and researching . Dear readers and followers, without you I am none. I always try to write different topics. I don't disappoint you to get new concepts and thoughts. As no scientist or researcher have not still developed a theory for the global warming and climate change. So I choose this topic to develop a model of global warming. Collecting necessary information from NASA and the World Bank Data Group about global warming and climate change , Environment Canada, Canada's national environmental agency for how much oxygen a tree produce per year , NC University for how much a tree absorb carbon dioxide per year I am trying and trying for a long time to develop a theory of global warming with my little knowledge. I am grateful these organizations. So if you find any mistake please inform me by email.

My wife and I are trying to educate our students with qualified teachers. They come from various economic groups but all the students of Notre Damian School get equal opportunity to learn. We hope ,we will create future Newton, Einstein, Socrates, Pythagoras and so on. If some one wants to help me to raise the fund for Notre Damian School I shall be grateful to him or her. Thanks for reading this section.

Mohammad Abu Shahadat

 08 August 2019

Dhaka, Bangladesh

Introduction

At present the global warming and climate change is the most talked topic in the world. We know that global warming means the increasing average global temperature. Various scientists and researchers give the reasons, effects and probable solution of this global problem in literature manner. But it is a matter of sorrow that no scientist or researcher develop a theory of global warming and climate change. It is I who am a student of welfare economic system and practitioner Ecophysics (Economics + Physics) making a mathematical model for the global warming. By this model we can calculate the global warming, calculate the amount CO_2, minimize the amount CO_2 (Carbon dioxide) for the green world, determine the amount of other greenhouse gases, determine the time when we will get the dreamy healthy world and find out the time when we control the weather of the world.

I have calculated one degree Celsius global warming is equal to (*48184380* + *a*) kiloton CO_2 and 1 kiloton CO_2 is equal to 5.42 kiloton O_2 in the context of absorbing by a tree. **One degree Celsius Global warming is equal to $66.02328T^{-1}$ degree adjacent angle**. In micro sense I have made the Global Warming Micro Model and then in macro sense I have established the Global Warming Macro Model. I believe that men have the power to control the weather of the world. It is our sacred duty to protect other animals and look after the earth like our sweet homes.

I think today's world is like a poor debt family in CO_2O_2 cycle. If CO_2 is Expenditure and O_2 (Oxygen) is income then the world's income (O_2) is less than CO_2. This is a problem and the problem is too large to imagine.

To develop this model I have used velocity, acceleration and differential. In micro sense, to solve the problem we have to increase O_2 and at the same time we have to minimize CO_2. In macro sense, we have to minimize greenhouse gases (CH_4, NO_2 and other green house gases) like compound interest loan and also minimize thermal energy which created by friction . This thermal energy is always ignored by scientists and researchers. I think this is a small matter but significantly big problem.

Fundamental Concepts of Global Warming Model

Global warming depends on the proportion of two groups elements. They are the increasing elements and the decreasing elements. The increasing elements are CO_2, CH_4 (Methane) , Nitrous Oxide and other greenhouse gases. The decreasing elements are O_2 and human's propensity to save the earth. If the value of proportion of two groups is more than one , the global warming increases and if the value of the proportion is less than one ,the global warming decreases.

The relationship between the global warming and the increasing elements (CO_2 and other GHG) is positive but there is negative relationship between the global warming and the decreasing element(O_2). Oxygen is only hero elements to protect the earth like a hero protects a heroine from evils. As trees can absorb CO_2 which is the most responsible element for global warming and release O_2 for animals living. Besides, more than sufficient trees can help human beings control CH_4, Nitrous Oxide and other greenhouse gases. So I concentrate on CO_2O_2 cycle which is not found any other planet I think. Many scientists believe that after 2030 many animals will be extincting for global warming. So it is important to know how much O_2 require against global warming and climate change, how much CO_2 should minimize from atmosphere to save the world, how many trees should have for living in the world, how to control weather for protecting against natural disasters. It is high time we did something for protecting animals including human beings from extinction. I am trying to save the loving world by developing theory, promoting it all over the world to apply it for the peaceful world of our next generation.

Global Warming Micro Model

" The more the value of the proportion of the efficiency of CO_2 emissions and the efficiency of O_2 production increases, the more the value of multiplier of global warming will increase and inversely decrease. When the value of the proportion is one, the Average Global Temperature will be constant in the context of CO_2O_2 cycle.. "

$w = c \ (x)^{-1}$ where, $c > 0, x > 0$ and if $w > 1$ then the global warming exists

$\theta = w \ (90 / 100)$ degree adjacent angle

Gw = Global warming = $\theta / 66.02328T^{-1}$ °C

GW indicates the Average Global Temperature or Global Warming

c = the efficiency of CO_2 emissions

= $(CO_2\,f - CO_2\,i\,) / CO_2\,f$

= $\Delta CO_2 / CO_2$ [CO_2f = final CO_2 and $CO_2\,i$ = initial CO_2]

x = the efficiency of O_2 production

= $(O_2\,f - O_2\,i\,) / O_2\,f$

= $\Delta O_2 / O_2$ [O_2f = final O_2 and $O_2\,i$ = initial O_2]

we get c by burning fossil fuels (i.g. coal, oil and gas) and x by planting trees and the number of trees in this world.

In global warming level $c > x$. The world people face this great problem now.

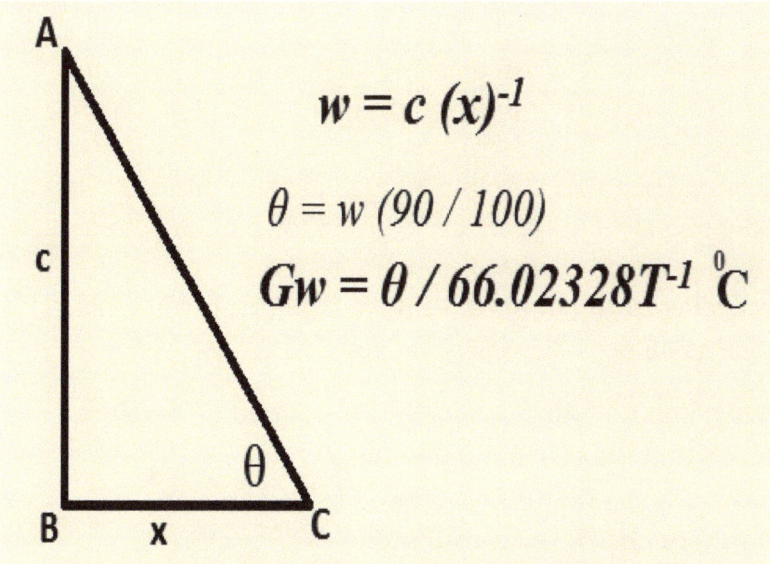

$$w = c\,(x)^{-1}$$

$$\theta = w\,(90 / 100)$$

$$Gw = \theta / 66.02328T^{-1}\ {}^{0}C$$

Figure: Global Warming Micro Model

In the figure , ABC is a right angled Triangle. Angle ABC = 90°, Angle ACB = θ, AB = perpendicular, BC = Base and AC = Hypotenuse

In 2013 the amount of CO_2 = CO_2i was 35837591 kiloton and 1n 2014 the amount of CO_2= CO_2f was 36138285 kiloton according to The World Bank Data Group.

c = (36138285 – 35837591) kt / 36138285 kt

= 300694 kt / 36138285 kt

= 0.00839046 kt

Therefore, c = ΔCO_2 / CO_2 = 0.00839046 kt

In 2013 the amount of forest land was 40057482 square kilometer and 1n 2014 the amount of forest land was 40024403.3 square kilometer according to The World Bank Data Group.

We know that a matured leafy tree produces 260 pounds O_2 in a year and this will do enough for two men according to Environment Canada, Canada's national environmental agency. They said," "On average, one tree produces nearly 260 pounds of oxygen each year. Two mature trees can provide enough oxygen for a family of four."

According to NC State University , "a tree can absorb as much as 48 pounds of CO_2 per year."

If we combine two concepts in the context of a tree, we get, 48 pounds CO_2 = 260 pounds O_2

1 pound CO_2 = (260 / 48)pounds O_2

= 5.42 pounds O_2 [approximately]

Therefore , 1 kiloton CO_2 = 5.42 kiloton O_2 [approximately]

That is, If we want to remove 1 kt CO_2 , we need 5.42 kt O_2. [appr.]

Suppose , T matured leafy trees are in one square kilometer

In one square kilometer,

T produces *(260 * T)/ 2204.6 kt O_2* [since 1ton = 2204.6 pounds]

= 0.12T ton O_2

So, one square kilometer forest land produces *0.12T kt O_2*

In 2013 the world produced the amount of *O_2 = O_2i*

= 40057482 * 0.12T ton O_2

= 4806897.84T ton O_2

= 4806897.84T / 1000 kt O_2 [since 1 kiloton = 1000 ton]

= 4806.89784T kt O_2

In 2014 the world produced the amount of O_2 = O_2f

= 40024403.3 * 0.12T ton O2

= 4802928.396 T ton O_2

= 4802928.396 T / 1000 kt O_2 [since 1 kiloton = 1000 ton]

= 4802.928396T kt O_2

ΔO_2 = (4802.928396T - 4806.89784T) kt O_2

= -3.969444T kt O_2

To save the earth for existing of animals we should increase the amount of oxygen but we cut down trees for our comfortable in exchange of our slowly but surely dead of our future generation.

Negative value of *ΔO_2* is the evidence of that word.

Avoiding negative sign we get *ΔO_2 = 3.969444 kt*

x = 3.969444T / 4802.928396T kt O_2

= 3.97T / 4802.93 T kt O_2

= 0.00082658T kt O_2

= 0.00082658T kt / 5.42 kt CO_2 [appr.] [since, 1kt CO_2 = 5.42 kt O_2]

= 0.0001525T kt CO_2

We, the inhabitants of the earth, make c = 0.00839046 kt CO_2

and *x = 0.0001525T kt CO_2*[approximately]

We have to increase the value of *x* more than *c*. But we have increased *c* more than necessity. As a result, the extinction of animals including human beings count down.

$w = c/x =$ 0.00839046 kt CO_2 / 0.0001525T kt CO_2

$= 55.0194T^{-1}$ kt CO_2

$w = 55.0194T^{-1} * 90$ degree / 100

$= 49.51746T^{-1}$ degree adjacent angle

According to NASA in 2014 the global temperature was 0.75 degree Celsius

0.75 degree Celsius global warming $= 49.51746T^{-1}$ degree adjacent angle

Or, 1 degree Celsius global warming

$= 49.51746T^{-1}$ degree adjacent angle / 0.75

$= 66.02328T^{-1}$ degree adjacent angle

So, one degree Celsius Global warming $= 66.02328T^{-1}$ degree adjacent angle

When the value of adjacent angle is one, we get the balance of $CO_2 O_2$

The more the value of *w* increases , the more the earth will unsuitable for living place and the more animal will extinct.

In balance level, c / x = 1

Or, $\Delta CO_2 / \Delta O_2 = 1$

Or, $\Delta CO_2 = \Delta O_2$

If ΔCO_2 is not equals to ΔO_2 , it means it does not establish balance.

If ΔCO_2 is greater than ΔO_2 , it means there is global warming .

If ΔCO_2 is smaller than ΔO_2 , it means the earth is suitable for animals.

The velocity of *Gw* is increasing because the velocity of ΔCO_2 is increasing and the velocity of ΔO_2 is decreasing.

As a result, the velocity of ΔCO_2 is more than the velocity of ΔO_2 and the velocity of VGw is increasing That is ,

$VGw = (Gwf - Gwi) / Time$,

$V\Delta CO_2 = (\Delta CO_2 f - \Delta CO_2 i) / Time$

$V\Delta O_2 = (\Delta O_2 f - \Delta O_2 i) / Time$

Therefore , $VGw = V \Delta CO_2 / V\Delta O_2$

the acceleration of Gw is a non-uniform acceleration because the value of CO_2 , O_2 and other greenhouse gases are variable.

In my global warming micro model I would not like to discuss the other Greenhouse gases such as Nitrous Oxide (N_2O), Mithene (CH_4),Chlorofluorocarbon (CFC_{12}), Hydrofluorocarbon-23 (HFC_{23}), Sulfer Hexa Fluoride (SF_6), Nitrozen Trifluoride (NF_3) because O_2 can not absorb these gases but by increasing O_2 it may reduce other greenhouse gases. So we have to minimize CO_2 on the priority basis.

According to NASA ,

In 2014 Average Global Temperature (Gw) = 0.75 degree Celsius

Average Global Temperature in 2000 according to NASA = 0.40 degree Celsius

The velocity of global warming from 2000 to 2014

= (0.75 – 0.40) / 14

 = 0.025 degree Celsius

The velocity of CO_2 from 2000 to 2014

= (36138285 kt – 24689911 kt) / 14

= 817741 kty^{-1}

The velocity of forest land from 2000 to 2014

= (40024403.3 km^2 - 40556022.3 km^2) / 14

= - 37972.786 km^2y^{-1}

Every year we lose approximately 37972.786 km^2y^{-1} forest land

Avoiding negative sign we get ,

the velocity of forest land is 37972.786 km^2y^{-1} .

Every year we lose approximately 37972.786 km^2y^{-1} forest land.

To manage deficit O_2,
We have to plant more trees.
We have to use renewable energy.
We have to avoid furniture made of wood.
Government should take Carbon tax from the factories which use fossil fuels and the individual who uses fossil fuel for driving personal car.
After managing deficit O_2 within 20 years we will reach the green world, I think. But we have not sufficient time to turn our beloved world into the green world within 20 years. We, the inhabitants of the world, are not united for minimizing carbon dioxide and increasing tree plantation. The United Nations sometimes fails to establish harmony between two conflict countries. Sometimes the United Nations remains silent to control the capitalist. However, We , the inhabitants of the earth, may vary from colour to colour, race to race, religion to religion but we all can unite for saving the world against global warming and climate change. To protect the earth against global warming we have to work unitedly by applying this theory, I think.

Calculation of how much CO_2 is responsible for one degree global warming

According to NASA in 2014,

the Average global temperature was 0.75 degree Celsius

And according to the World Bank Carbon emissions was 36138285 kiloton

So , we get,

0.75 degree Celsius Gw = 36138285 kiloton CO_2 + GHGo

Or, 1 degree Celsius Gw = *(36138285 kt CO_2 + GHGo) /* 0.75

Or, 1 ^oC Gw = (36138285 kt CO_2 / 0.75) + (GHGo / 0.75)

Or , 1 ^oC Gw = 48184380 kt co_2 + a

[let, *a* =GHGo / 0.75]

As O_2 can absorb CO_2. *a* can be minimized by human behaviors. So in micro sense, Global warming multiplier depends on the proportion of *c* and *x*

One degree Celsius Global Warming is equal to 48184380 kt CO_2 + *a*

The Global Warming Macro Model

In macro sense,the Creator gives human beings the power to control the atmosphere because everything of the universe is made for human beings. Global Warming depends on human activities. Two types of coefficients are created by human activities . they are increasing elements of global warming and decreasing elements of global warming.

If the proportion of the sum of the marginal propensity of increasing elements is equal to the the sum of the marginal propensity of decreasing elements , the changing rate of Global Warming will be constant and establish the green world.

The global warming Macro function,

$GW = f(P) = f (CO_2, O_2, Re, GHGo, AntiGHGo ,TEc)$

Where, GW = Global Warming , P = Population, CO_2= Carbon dioxide, O_2 = Oxygen, Re = Renewable Energy, GHGo = CH_4 ,N_2O and other greenhouse gases, TEc = Thermal Energy caused by friction, AntiGHGo = Action against GHGo

$GW = f(P) = hP$

Therefore, $dGW / dP = h > 0$

where, P = Population of the world , h = the marginal propensity of human being's behavior

The global warming macro equation,

$GW = f(CO_2) + f (GHGo) - f (O_2) – f(Re) – f(AntiGHGo)+ f(TEc)$

Part of $GW = f (CO_2) = kCO_2$

Therefore, $dGW / dCO_2 = k > 0$

Where, c = the marginal propensity of CO_2 emissions

Part of $GW = f(O_2) = -sO_2$

Therefore, $dGW / dO_2 = -s \quad < 0$

Where, x = the marginal propensity of O_2

Part of $GW = f(Re) = -r\,RE$

Therefore, $dGW / dRE = -r \quad < 0$

Where, r = the marginal propensity of using Renewable Energy

Part of $GW = f(GHGo) = g\,GHGo$

Therefore, $dGW / dGHGo = g \quad > 0$

Where, g = the marginal propensity of other greenhouse gases

Part of $GW = f(Tec) = e\,TEc$

Therefore, $dGW / dTEc = e \quad > 0$

Part of $GW = f(AntiGHGo) = -y\,AntiGHGo$

Therefore, $dGW / dAntiGHGo = -y \quad < 0$

Where, y = the marginal propensity of antiGHGo

The global warming macro equation,

$GW = k\,CO_2 + g\,GHGo - s\,O_2 - r\,Re - y\,AntiGHGo + e\,Tec$

Or, $\Delta GW / \Delta(CO_2 + O_2 + Re + GHGo + AntiGHGo + TEc) = k + g - s - r - y + e$

Or, $\Delta GW / \Delta P = h = \tan\theta$

If h = 45 degree then there will establish ecological balance

But it is a matter of sorrow that today's world face $h = \tan\theta$ > 45 degree . For this reason we face global warming.

When $h = \tan\theta = 45$ degree, then $(k+g+e)/(r+s+y) = 1$

When, tan θ > 45 dgree, then *(k+g+e) / (r + s + y) > 1*
When, tan θ < 45 dgree, then *(k+g+e) / (r + s + y)   < 1*

Here, *(k + g + e)* is the sum of k,g and e
      (r + s + y) is the sum of r,s and y
So we all, the people of the world, have to work for increasing the value of (r + s + y) and decreasing the value of (k + g + e).
Now the world faces the following diagram,

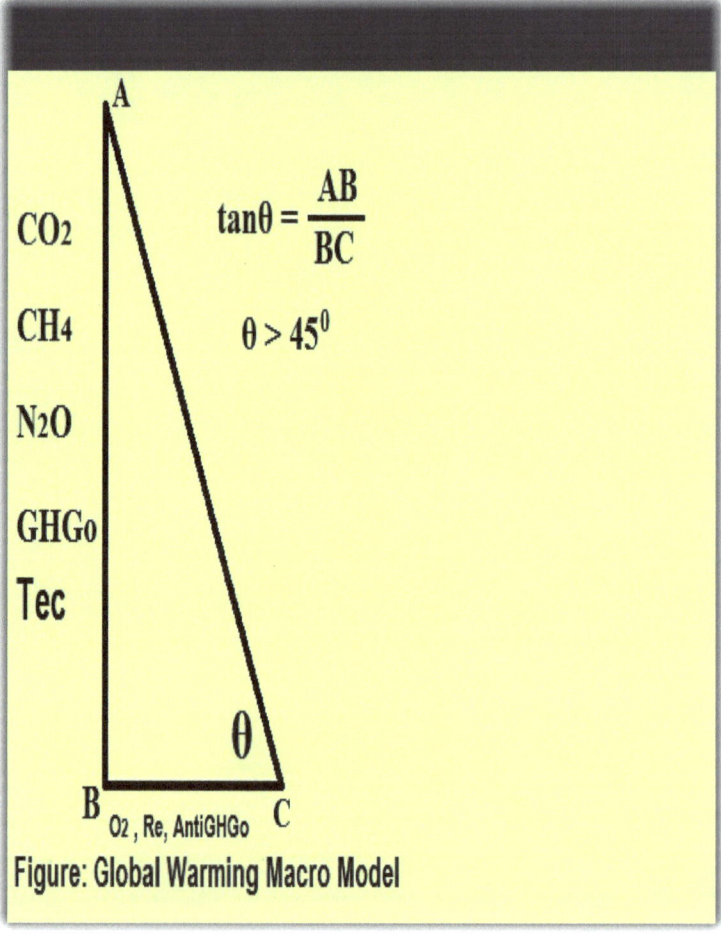

Figure: Global Warming Macro Model

In the figure, ABC is a right angled triangle. Angle ABC = 90°,

angle ACB = θ

AB / BC = $\tan\theta$ = h = (k + g + e) / (r + s + y)

The 21th century of people face the gigantic problem which is called the global warming. For the global warming the world people are seeing how sea water level is increasing, polar ice is melting, many islands is sinking, many animals are disappearing and seeing various natural disasters. Suppose , you want to boil a kettle of water by heating. What will happen when the water of the kettle boils ? You can see that water level of the kettle is rising. Like the kettle our earth is heating by burning fossil fuels, using fertilizer and using various types of glass for building high rise tower, using automobiles .This heat is the main cause of increasing global temperature and increasing sea water level. When water heats up, it increases. So when the ocean warms, sea level rises. Polar ice is melting for global warming and water created by melting water runs into the ocean and thus sea water is rising gradually and it helps methane (CH_4) emit.

Many of us believe that many islands around the world are slowly but surely submerged by 2050 because of rising sea water level. This will happen only for global warming.

Some Functions for Global Warming

According to the macro model, there are some functions for global warming. I would like to discuss these functions. The first function is called carbon dioxide - global warming function.

The carbon dioxide – global warming function

Global warming depends on mainly CO_2. There are positive relationship between G_w and CO_2.
The more CO_2 increases the more global warming increases and inversely decreases.
G_w is dependent variable and CO_2 is an independent variable.
So the function is,

$Gw = f(CO_2)$

The equation of the global warming – carbon dioxide is,

$Gw = k\, CO_2$

Therefore, $\Delta Gw / \Delta CO_2 = k$

The Global Warming-Oxygen Function

There is an inverse relationship between Gw and O_2. The more the amount of oxygen increases the more the amount of Gw decreases.

The function of global warming – Oxygen is ,

$Gw = f(O_2) = -s\, O_2$

Therefore, $\Delta Gw / \Delta O_2 = -s$

To minimize global warming we have to increase the quantity of forest land .Trees take CO_2 and give O_2 to maintain ecological balance. But at present the lower amount of O_2 do not perfectly absorb the higher amount of CO_2 because of human activities. We have to plant trees here and there where we find open space I think every man needs at least 226 trees to exist in the world. So we have to plant at least 226 trees per capita. By planting trees we can save many human lives and wild lives and maintain ecological balance.

The function of Carbon dioxide minimization in atmosphere by O_2

Naturally 5.42 unit oxygen is equal to 1 unit CO_2.

$$CO_2 = f(O_2) = t\,O_2$$

Or, $CO_2/t = O_2$

Or, $CO_2\,(t)^{-1} = O_2$

Or, $(t)^{-1} = O_2/CO_2$

Since, $CO_2 > O_2$ and in global warming level CO_2 is positive

By planting trees we will get $t = 5.42$

When we get $t = 5.42$, we will minimize $CO2$

At present t is less than 5.42. We have to face natural disasters and climate change.

The G7 (Group of Seven) countries like USA, Canada, Japan, England, France, Germany, Italy and developing counties like Chine, Brazil, India, Australia , Rasia etc use fossil fuels (i.e. coal, oil, natural gas) for increasing GNI (Gross National Income) and produce a large number of Carbon dioxide which is responsible for climate change. Most of them are unwilling to reduce the use of fossil fuels.

Due to deforestation the present number of trees can not normally absorb CO_2 from the air. If we have to eat 50 kilogram rice without any time interval, what will happen ? Just imagine. Same condition exists for the number trees of the world in the context of absorbing CO_2. Rather trees lose their normal power to absorb CO_2 and release poor oxygen.

Renewable Energy- Global Warming Function

The more the users of renewable energy increase, the more the global warming decrease. There is a negative relationship between Re and Gw. We have to use renewal energy i.g. hydro power, biomass, solar panel, wind power, biofuel etc.

$G_W = f (Re) = -rRe$
$\Delta G_W / \Delta Re = - r$ where, $r > 0$, it indicates the marginal propensity to use renewable energy.

Greenhouse Gases – Global Warming Function
The more GHGo increases, the more the global warming increases. There is a positive relationship between G_W and GHGo
$G_W = f(GHGo) = gGHGo$
$\Delta G_W / \Delta GHGo = g$ where, g is the marginal propensity of GHGo

Global Warming -Thermal Energy caused by friction Function
The more the thermal energy caused by friction increases, the more the global warming increases. There is a positive relationship between GW and Tec
$G_W = f (Tec) = e\, TEc$
$\Delta GW / \Delta TEc = e$ where, e = the marginal propensity of thermal energy

AntiGreenhouse Gases – Global Warming Function
The more AntiGHGo increases, the more the global warming decreases. There is a negative relationship between G_W and AntiGHGo

$G_W = f(AntiGHGo) = y\, AntiGHGo$
$\Delta G_W / \Delta AntiGHGo = y$ where, y is the marginal propensity of AntiGHGo

Who Are Responsible for Global Warming

Human beings create global warming. Nature have no power to create global warming. We, the human beings, are responsible for increasing global warming. We are decreasing the number of trees per square kilometer for making furniture, building living places and manufacturing various products from trees. As a result , we get low amount of oxygen from air but we produce carbon dioxide more than necessary in air. The lower capital than necessary begets the lower production. The lower production than necessity are responsible for inflation. Like this way we begets global warming by getting lower amount of O_2 and releasing more CO_2 in air. How many trees should need for a square kilometer land according to the density of population is unknown to us. I think it is time to calculate the number of trees in a square kilometer by research. It is not necessary to travel another planet like The Mars. It is necessary to find out how many trees we need a square kilometer land to get enough oxygen. Lack of oxygen we produce Carbon dioxide more than necessary.

How to Reduce the Global Warming

To reduce global warming we have to increase the marginal propensity to plant trees and use renewable energy so that CO_2 can not increase. If we use more fossil fuels we will increase more CO_2. So we have to minimize CO_2. There is positive relationship between CO_2 and global warming. If CO_2 increases, global warming will increase. We know that a matured leafy tree produces 260 pounds O_2 in a year and this will do enough for two men according to Environment Canada, Canada's national environmental agency. They said," "On average, one tree produces nearly 260 pounds of oxygen each year. Two mature trees can provide enough oxygen for a family of four."

According to NC State University , "a tree can absorb as much as 48 pounds of CO_2 per year."

If we combine two concepts, we get, 48 pounds CO_2 = 260 pounds O_2

1 pound CO_2 = (260 / 48)pounds O_2 = 5.42 pounds O_2

Therefore , **1 kiloton CO_2 = 5.42 kiloton O_2**

That is, If we want to remove 1 kt CO_2 , we need 5.42 kt O_2.

In 2018 the number of world population is 7.594×10^9 according to The World bank Data Group.

O_2 needs in 2018 for the whole population

$= 7.594 \times 10^9 \times 130$ pounds

$= 987.22 \times 10^9$ pounds

$= (987.22 \times 10^9) / 2204.6$ ton

$= 447800054.43$ ton O_2

$= (447800054.43 / 1000)$ kt O_2

$= 447800.054$ kt O_2

Suppose, for all animals O_2 need in 2019 ,

4 times x human beings required O_2

$= 4 \times 447800.054$ kt O_2

$= 1791200.22$ kt O_2

$= (179200.22 / 5.42)$ kt CO_2

$= 330479.74$ kt CO_2

But in 2014 we, the inhabitants of the earth, produced 36138285 kt CO_2 according to the World Bank Data Group. This amount is increasing day by day. In 2019 the earth need 330479.74 kt CO_2 but we produced 36138285 kt CO_2 in 2014 by cutting down trees, using fossil fuels (i.g. coals, gases, oils) . Extra amount we produced 5 years ago (36138285 − 330479.74) = 35807805.26 kt CO_2 which was **108.35 times more.** The gigantic problem is this amount CO_2.

On the other hand, the world has only 39958245.9 square kilometer forest land. How many trees are there in a square kilometer forest land ? We don't know. But we know we are minimizing forest land by cutting trees for making furniture, using industrial purposes etc.

Suppose , T matured leafy trees are in one square kilometer. The number of trees in the world in 2016 is 39958245.9 x T trees.

They produced 3995845.9 T (260 / 2204.6) ton

$= 3995845.9 \ T \times 0.12$ ton

$= 479501.508 T$ ton O_2

$= 479501.508T \ / \ 1000 \ \ \text{kt} \ O_2$

$= 479.502T \ \ \text{kt} \ O_2$

The Green World

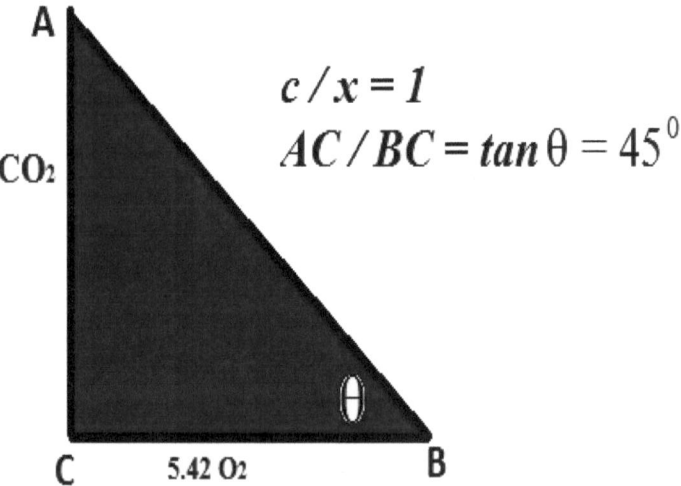

$$c \ / \ x = 1$$
$$AC \ / \ BC = tan \ \theta = 45^0$$

Figure : The Green World

In green world level, $CO_2 \ / 5.42 \ O_2 = 1$

In the figure , According to the right angle triangle ABC,

$\tan \theta = AC \ / \ BC = CO2 \ / \ O2 \ = 1 = 45$ degree

In the green world level, everywhere we will find trees and trees. The world will seem to be a big jungle. The inhabitants of the world will overcome natural disasters and live happy naturally. The proportion of the marginal propensity of using fossil fuels and the tree plantation will same, there have no extinction of any animal. There will establish ecological balance.

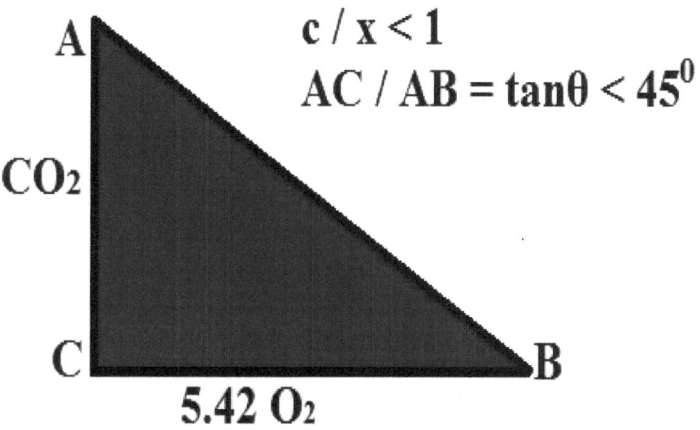

$$c / x < 1$$
$$AC / AB = \tan\theta < 45^0$$

A

CO_2

C   5.42 O_2 B

Figure: The Healthy Dreamy World

When , the world reach $CO_2 / 5.42\ O_2 < 1$,
then the world will be called Dreamy Healthy World. I think every man needs at least 226 trees to exist in the world. So we have to plant at least 226 trees per capita. By planting trees we can save many men and women and wild lives and maintain ecological balance.

.

How Can We Turn the Present World into the Green World

The present world will be unsuitable for living by 2050. Many animals will extinct by this time. So it is high time we, the people of the world, increased the marginal propensity to produce oxygen and decreased the marginal propensity of carbon emissions so that we can turn the present world into the green world. According to my model when we reach $c = x$, we will make the existing world the green earth.

For increasing the number trees we have to take the following steps.

The marginal propensity to plant trees should be greater than the marginal propensity to cut down trees.

that is, $p > d$ Where, $p, d > 0$

p = the marginal propensity to plant trees ,

d = the marginal propensity to cut down trees

if $p > d$ then the world people can increase the number of trees . that means $x = p - d$ where, $x > 0$

If p < d , then the world people can decrease the number of trees for their extinction.

That means $x = p - d$ where $x < 0$ which we the world people have created. For this we are facing global warming. To save the world against global warming we should increase the value of x .

The United Nations, World Bank and climate related organizations should encourage the world people to plant trees. The botany experts should research for knowing how much O_2 (oxygen) produce per tree. We should plant the tree which produces the largest amount of oxygen .

To save the world against the global warming we should decrease the value of w . To minimize the amount of carbon dioxide we have to use more renewable energy than fossil fuels.

If the marginal propensity to use fossil fuels is less than the marginal propensity to use renewable energy , we will decrease the value of w.

Besides ,

We should be aware of using the things which produce the greenhouse gases.

The Botany experts should find out how many trees need for one square kilometer forest land and find out the tree that produce more oxygen than normal tree.

We always remember that space travelling is less important than tree plantation. The earth is the only place for animals living because CO_2O_2 cycle exists only this planet and the central point of the universe is here . I have discovered the central point of the universe by my Universal Model. Finally I would like to say " First save the earth against global warming then research another planets.

Analysis of the collected data

Year	P (Billion)	ΔP	Forest Area (Sq.km)	ΔF	CO_2 Emis. (Kiloton)	ΔCO_2
1999	6.035	-	40628689.7	-	24059187	-
2000	6.115	0.080	40556022.3	-72667.4	24689911	630724
2001	6.194	0.079	40510303.3	-45719	25276631	586720
2002	6.274	0.08	40464584.1	-45719.2	25646998	370367
2003	6.353	0.079	40418865.2	-45718.9	27047792	1400794
2004	6.432	0.079	40373146	-45719.2	28393581	1345789
2005	6.513	0.081	40327427	-45719	29490014	1096433
2006	6.594	0.081	40293287.6	-34139.4	30568112	1078098
2007	6.675	0.081	40259147.7	-34139.9	31180501	612389
2008	6.758	0.083	40225008.7	-34139	32181592	1001091
2009	6.841	0.083	40190869.1	-34139.6	31891899	-2829693
2010	6.923	0.082	40156729.7	-34139.4	33472376	1580477
2011	7.004	0.081	40123639.2	-33090.5	34847501	1375125
2012	7.087	0.083	40090560.5	-33078.7	35470891	623390
2013	7.171	0.084	40057482	-33078.5	35837591	366700
2014	7.256	0.085	40024403.3	-33078.7	36138285	300694
		ΣΔP = 1.221		ΣΔF = -604286.4		ΣΔCO2 = 9539098

Data source :

World Bank Data Group (Population, Forest Area and CO2 Emissions).

Here, P = Population,

ΔP = current year – previous year,

ΔF = Current year – previous year ,

ΔCO_2 = Current year – previous year

The average value of ΔP from 2000 to 2014

= ΣΔP / n = 1.221 / 15 = 0.814

The average value of ΔF from 2000 to 2014 = $\Sigma\Delta F$ / n = - 604286.4 / 15 = - 40285.76 sq.km.The average value of ΔCO_2 from 2000 to 2014 = $\Sigma\Delta CO_2$/ n

= 9539098/15 = 635939.87 kt

Comment: there is positive relationship between Population and Carbon dioxide but negative relationship between population and forest area. So lack of sufficient oxygen the world temperature is increasing gradually.

The velocity of CO_2 emissions from 2000 to 2014 (VCO_2)

= (36138285 – 24689911) / 15

= 763224.93 kt

The velocity of forest area from 2000 to 2014(VF)

= (40024403.3 – 40556022.3) / 15

= - 531619 square land

Suppose, the number of trees per square km = T

According to Canada Environment Agency ,

 1 matured tree produces O_2 = 260 pounds.

According to them,

A tree produces O2 = 260 / 2204.6 = 0.12 ton

We know that 1000 ton = 1 kiloton

T trees produce O_2 per square kilometer = 0.12T ton

The velocity of O_2 from 2000 to 2014 (VO_2)

= VF x 0.12T ton = - 531619 x 0.12 T ton

 = - 63794.28 T ton

= - 63794.28 T / 1000 = - 63.79T kt

Every year we lose O_2 63.79 T kt (approximately)

So we, the inhabitants of the earth, should produce 2 times of 63.79 T kt O_2 per year for saving the world.

The velocity of the global warming from 2000 to 2014 (VGw) = (0.75 – 0.40) / 15 °C = 0.023 °C

The velocity of Population growth = $(7.256 \times 10^9 - 6.115 \times 10^9) / 15$ $= 1.141 \times 10^9 / 1$ $= 76066666.67$

From the above discussion , It can be said that in nature there are negative relationship between global warming and forest areas which produce O_2 for maintaining ecological balance. **So only through increasing forest area global warming can be minimized.**

O_2 inhaled from air in 2014

= (Total forest area x 0.12 T) / (7.256×10^9) ton per person

= $(40024403.3 \times 0.12 T) / (7.256 \times 10^9)$ ton per person

= 4802928.396 T / 7256000000 ton per person

= 0.00066 T ton / person

$CO2$ exhaled in air in 2014 year

= 36138285 / (7.256×10^9) kt/ person

= $(3.6138285 \times 10^7) / (7.256 \times 10^9)$

= 3.6138285 / 7.256 X 10^2

= 3.6138285/ 725.6

= 0.0049805 kt/ person

= 4.9805 ton/ person

According to NC State University , a tree can absorb as much as 48 pounds of CO_2 per year.

So , 48 pounds = 48 /2204.6 ton = 0.022 ton (approximately)

The number of trees is needed for absorbing CO_2

= 4.9805 / 0.022 trees per person

= 226 trees (Approximately)

A human being should plant 226 trees (approximately) for saving the world against climate change and global warming.

According to World bank Data Group, in 2000

Carbon emissions 24689911 kt,
Methane(CH_4) Emissions 6480650 kt (equivalent CO_2),
Nitrous oxide emissions 2920510 kt (equivalent CO_2)
According to NASA, Global Warming 0.40° in 2000
So, Gw = CO_2 + CH_4 + N_2O + GHGo
0.40° Gw = (24689911 + 6480650 + 2920510) kt + GHGo
Or, 0.40° Gw = 34091071 kt CO2 + GHGo
Or, 1° Gw = 85227677.5 kt CO2 + (GHGo / 0.40°)
Or, 1° Gw = 85227677.5 kt CO2   [(GHGo / 0.40°) ≈ 0]
So, 1° Gw = 85227677.5 kt CO2 (approximately)

It is high time we planted at least 226 tree plants wherever we like for saving our next generation like my only son Shayan against global warming and climate change.

Result of the Theory

In micro model,

$w = c\ (x)^{-1}$ where, $c > 0$, $x > 0$ and if $w > 1$ then the global warming exists

$\theta = w\ (90 / 100)$ degree adjacent angle

Gw = Global warming = $\theta / 66.02328T^{-1}$ °C

1 pound CO_2 = 5.42 pounds O_2 [approximately]
1 kiloton CO_2 = 5.42 kiloton O_2 [approximately]
If we want to remove 1 kt CO_2 ,
we need 5.42 kt O_2. [approximately]

Suppose , T matured leafy trees are in one square kilometer one square kilometer forest land produces $0.12T$ kt O_2

one degree Celsius Global warming = 66.02328T $^{-1}$ degree adjacent angle

one °C global warming is equal to (*48184380* + *a*) kiloton CO_2

Macro model,

GW = hP

$\quad$ = KCO_2 + g GHGo + eTEc − sO_2 − r RE − y AntiGHGo

Global Warming depends on population behavior.

To remove global warming every person needs to plant 226 trees for saving the world and our future generation. Otherwise, you may become the richest person in the world but you will not be able to save your life against global warming.

I think , by applying this model we, the inhabitants of the world, can minimize 0.30 °C global warming within 5 years and within 20 years we will get the healthy dreamy world.

The Universal Model

The universe has a central point. I have found it. If the central point's charging power is zero, all the things of the universe lose their gravity and collide one another. The radius of the Universe is 127r ly ....

The universal model will be published soon.

By this model we will know –

The central point of the universe?

Theory of the central point?

How to stop the all the things of the universe motion?

If the central point's power is zero, then what will happen?

The radius , area, circumference of the universe

Equation of the existence of the Creator?

Differences between the Big Bang and my Universal Model

Dear reader, thanks for reading this book. If you find any mistake, you are earnestly requested you to send me your comment.

mabu.shahadat@gmail.com

www.facebook.com/MAbuShahadat

www.ingramcontent.com/pod-product-compliance
Lightning Source LLC
Chambersburg PA
CBHW041117180526
45172CB00001B/293